THOUGHTS ON BEING A CHIEF PETTY OFFICER

BY

Lieutenant Edward Leo Semler Jr., USCG (Retired)

Copyright © 2017 by Edward Leo Semler Jr.

All rights reserved by the author.

First Edition: 2017

Library of Congress Control Number: 2017905951

ISBN: 978-0-692-87867-5

Printed in the United States of America

Cover picture and text by Edward Leo Semler Jr.

To my son Garrett

&

To all Chief Petty Officers

Table of Contents

Introduction	1
Leadership training	5
Know yourself	11
Chiefs Mess	27
Command Chief	41
You and your junior officer	47
Setting the example	55
Mentor your people	67
Life after Chief	75
Chiefs Creed	81
References	87
Notes	89
About the author	91

Introduction

After I retired from the U.S. Coast Guard I moved to rural Pennsylvania, then to Texas, far from the active military community and any type of military installation. So my interaction with anyone on active duty has been limited to catching up with old shipmates every once in a while via phone and social media. And a majority of the people who have contacted me are junior shipmates I sailed with. They're now advancing to chief petty officer and want my "words of wisdom."

I'm always very flattered that they take the time to reach out to me for my thoughts on leadership. And as I compile my thoughts on the matter, so many things come to mind that I want to share with these new chiefs. More than I can fit on the single piece of paper they send me. Through the years I came up with a canned response, something along the lines of remember the good chiefs you worked for and emulate what they did, but also remember the bad chiefs and don't repeat their mistakes.

As broad and true as that bit of advice may be, I felt I was doing these young chiefs an injustice with just that statement. So I decided to put together my thoughts on what I believe makes a chief. And that's exactly what this book is: *my own personal thoughts* on being a well-rounded chief. It's based on my experiences, good and bad. And this book isn't just for the prospective or new chief. Senior and master chiefs may find it just as helpful – or at least familiar. And for the Navy chiefs reading this I think you'll find that a chief is a chief no matter what branch of the service you're in.

Well I guess your first thought is; what gives this guy's opinions on being a chief petty officer any credence. Good point. My dad use to tell me all the time about opinions and advice, "You can take it or leave it, it's up to you."

What I have to offer is experience. Experience as a chief on land and afloat. I made chief in 1994, senior chief in 1998, and master chief in 2001. I served in numerous all male and mixed gender chiefs messes and as a command senior and master chief.

If you're interested I wrote another book all about my career entitled *"Around the World; 25 years of service as an officer and enlisted man in the U.S. Army and U.S. Coast Guard."* I use several of the stories from that book in this one because I feel they're good real life examples of what you're going to encounter as a chief. All the stories you read here are true; at least as I remember them. And since they happened to me they're not that far removed from what you may experience as a chief, or already have.

What I'm going to lay out for you in this book are my thoughts and real life experiences which helped me grow as a chief. Keeping the topics short and to the point; my intention was to compile a sort of reference guide for you.

Was I always a good chief – no! And you probably won't be either. You'll have your good days and your bad days. But hopefully after reading this book you'll limit your bad days.

And if you only take away one thing from this book that helps you I'll consider it a success. Because when you're a better chief, it benefits everyone you interact with.

Leadership training

No matter what paygrade you hold you're more than likely going to be overseeing somebody. And every bit of leadership training or mentoring you can get will help you. Because when you join the service as an enlisted person you don't come in with the mindset of being a leader. You're enticed by your recruiter to come into a trade and learn a skill. The last thing you're thinking about is managing people and enforcing discipline. Officers are the exact opposite. They don't join with the expectation of learning a trade; they join to lead.

So leadership training and mentoring are very important to the enlisted person. But learning and retaining it can be difficult because people take in information differently. I'm one of those people who just have a tough time with classes and a structured setting. In fact I obtained my Associate and Bachelor degrees all through the college level entrance program (CLEP) exams, without taking a single traditional classroom course. I'm the type of person who needs hands on training. I learned most of my leadership skills watching people in leadership positions and observing where they succeeded and failed.

And if you learn better in a classroom setting great! Make sure you take advantage of every course available to you. And better yet if your unit or service offers a mentoring program get involved. In my opinion there is nothing better than personal guidance and advice.

I have attended numerous leadership schools throughout my 25 year military career. And I enjoyed them all because I always took something away from them; however small. Probably the most prestigious was the Chiefs Academy. But I didn't attend the Coast Guard or Navy Chiefs Academy, I attended the Air Force Senior Non-Commissioned Officers Academy. I say that not to insinuate that one is better than the other. But just to say that the military services cross train; which I think is a great idea.

As I previously mentioned, I don't retain information well in a classroom setting. So after 33 days of intense Air Force leadership training what influenced me the most was the movie "Twelve O'clock High."[1] The movie is set during World War II depicting a fictional Air Force Bomber Wing. If you watch it with a mindset on leadership and management you'll be surprised how well it is written for just that. And as you watch, it's also easy to find yourself identifying with any one of the characters; because it portrays a wide spectrum of military personalities in the movie.

The movie in a nutshell is about an overly friendly and concerned commanding officer. His command climate becomes lax as he overcompensates by loosening the rules in exchange for his men risking their lives on combat missions. The commanding officer is eventually relieved after having a nervous breakdown worrying about his men. In comes a new commanding officer who shapes up the command by enforcing the rules and regulations. Obviously not liked, but he gets results.

The way the commanding officers interact with their peers and subordinates is textbook leadership, or lack of it, on film. And even though this depicts an Air Force unit in the film, the leadership issues cross service boundaries. In our case just step it down to a new chief taking over a division. It plays out just the same.

I read a lot of World War I & II memoirs and autobiographies about enlisted and junior officers. I just find that first hand stories are a lot more interesting than those written or interpreted by a second person. And I find a recurring theme in all of them; sound leadership and authority get results. Subordinates feel and behave better when they have a leader they can trust and who administers consistent, unbiased discipline.

I have seen chiefs try to become well-liked by their subordinates. And when it all falls apart they wonder why? Because they became too friendly with their subordinates. Yea, that's a problem. As a chief you're not in a likability contest. I'm not saying you need to act like Captain Queeg in the movie "The Caine Mutiny,"[2] but you need to be in control, impartial, fair, and consistent.

Speaking of "The Caine Mutiny," I think it's another good leadership film. It starts out with a U.S. Navy vessel whose Commanding Officer is replaced due to normal rotation. Like in "Twelve O'clock High" the previous commanding officer was well liked and maybe a bit lax. The strict new Commanding Officer, Captain Queeg, wants more discipline from what he perceives as a lax crew, which immediately draws resentment. But unlike "Twelve O'clock High," this new commanding officer also has a major flaw: he is mentally unstable. Because of this his wardroom becomes dysfunctional and distrusts him. In the end, the executive officer relieves the commanding officer at the height of a typhoon when the ship is floundering and the commanding officer can't make a decision. The commanding officer, once he has regained his composure after the storm, accuses the executive and communications officer of mutiny - and they go to Court Martial. The communications officer fails to support the executive officer and lies under oath to distance himself from the executive officers decision to relieve Captain Queeg. But rightfully the two officers are eventually acquitted of the mutiny charge.

You have a pretty good chance of encountering people like Captain Queeg and the communications officer in the fleet; flawed, dysfunctional, and just plain bad officers and fellow chiefs. Don't believe me?

In 2016 the Navy found guilty at Court Martial 1 captain, 2 lieutenant commanders, 1 lieutenant, 1 ensign, 1 warrant officer, 7 senior chiefs, and 15 chiefs.[3] No Non Judicial Punishment data was available for 2016 - but in fiscal year 2015 the Navy relieved for cause 1 admiral, 9 captains, 9 commanders, 5 lieutenant commanders, 6 master chiefs, and 1 senior chief.[4]

And in fiscal year 2016 the Coast Guard court martialed 1 captain, 1 commander, 1 lieutenant, 1 warrant officer 1 master chief, 2 senior chiefs, and 7 chiefs. Military administrative action was taken on 1 captain, 4 commanders, 10 lieutenants, 1 lieutenant junior grade, 1 ensign, and 3 warrant officers. And relieved for cause were 2 warrant officers and 1 chief.[5]

And these are just the ones that got caught. So believe me when I tell you they are out there.

Hopefully you won't have to work for or with them, but what if you do? Will you be able to confront them? Go to a court martial like the executive officer under Captain Queeg and stick by your actions? Hopefully this book will help you prepare for such an encounter and keep you from becoming a statistic.

Know yourself

Recognizing your strengths and weaknesses will help you become a better chief. Leadership has its ups and downs which can drive a roller coaster of emotions that will affect your behavior. We're all human and have personality flaws. Hopefully not to the extent that I previously mentioned, but minor quirks. When you're in a stressful leadership situation - angry, and about to lose your temper - your real personality is going to come out, not the cool character you strive to project. You're not going to be thinking about leadership training and whether the situation requires a passive approach, aggressive approach, or whatever; you're just going to react.

Being honest with yourself and about your weaknesses will help you recognize when you are about to go off the rails, and hopefully enable you to control yourself and the situation. And if for some reason you do lose your composure, you should never let your pride keep you from apologizing. Yes, I said apologize. Remember your Chiefs Creed and the part about having humility. Setting the record straight with your subordinates will go a long way with them.

I was the main propulsion senior chief with twelve enlisted, E-2 to E-6 working for me. I had an agreement with them that if they worked after hours I would compensate them with extra liberty.

I had a hard working E-3 who was just out of boot camp. His home of record was about five hundred miles away and he wanted to go back and pick up his car. He sent a request up the engine room chain of command for special liberty, off the record, to go home. My E-6 came to me with his request, recommending I disapprove it because he thought the E-3 was too immature. I felt my E-6 was being too harsh and told him to let him go. I thought I had observed the E-3 pretty well in the short time he was aboard and his work ethic impressed me.

The E-3's home was a good eight hour drive south, but he assured me that he was catching a ride home with another shipmate going on official leave, was going to get his car, and come straight back. It all sounded pretty cut and dry. I told him he could have Friday, Saturday, and Sunday off since he didn't have duty. And to make sure he was back on board Sunday night for work on Monday when we would have all hands quarters. I was pushing it because legally I shouldn't let him go that far from the cutter and for that long. Sunday night and Monday morning rolled around with no sight of the E-3 and he missed quarters. Of course my E-6 was giving me the, "I told you so" look. I pulled up the E-3's personnel record from the ship's office and found his parents' phone number. I called the number and reached his mom and dad. I asked them, "Has your son been home?" And they said, "Yes, but only for a short while and he left." I asked, "Did he leave with his car?" And they said, "He doesn't own a car." As I talked to the parents it was clear that his son, my E-3, was a troubled kid. The father explained to me that they had pushed him into the service to straighten him out. Oh great, just what I needed to hear! Now I had to go to my boss and explain what had happened. Because I was the one who let the E-3 go, I needed to own up to it. My boss in turn had to go to the commanding officer and explain the matter. Not good for my boss, not good for me, and it sure as heck wasn't going to be good for the E-3.

Needless to say, the E-3 was listed as absent without leave (AWOL). Now the Coast Guard doesn't go and look for you if you go AWOL. They don't have the resources and pretty much hope you'll come back. Nothing is really done until after 30 days when you become a deserter. At that point you're placed on a report which will identify you if you're stopped by law enforcement for any reason. In which case you would come up as a deserter and they would detain you and hand you back over to the Coast Guard.

After about a week I received a call from the E-3 who was out running wild with his friends back home.

I was able to convince him that if he came back before 30 days were up he would be a lot better off and only face a Captain's Mast. If he waited over 30 days, he would be a deserter and go to the brig. He did come back and went to Captain's Mast for being AWOL and got the usual 30 days extra duty and 30 days restriction. I deservingly got a verbal reprimand.

I made two mistakes here; first, I really didn't know my E-3 as well as I thought and was putting way too much responsibility on him, which he couldn't handle. Second, I didn't listen to my E-6 who was in a position to know the E-3 better than me.

And regrettably this wasn't the first time this had happened to me. The incident I just explained happened when I was a senior chief. When I was a chief running an engine room on another cutter the exact same thing happened to me. So it was a flaw that I should have known I had. I had a bad habit of putting too much responsibility onto the shoulders of those who were not ready for it. I should have paid more attention to what I was letting this kid do. And I should have left the call up to my E-6 who knew the E-3 better than I did. And afterward I told the E-6 that he was right - and I should have listened to him. I also apologized to my boss for putting him in a bad situation.

As a chief you'll constantly be navigating through issues you need to be involved in, and even more that you need to steer clear of. Subordinates will always be trying to get you involved in their dramas. It's just a part of human nature. You control a lot of positional power and some people below you want to tap into it for personal gain.

We were underway and rather than being served in the chiefs mess, I was going through the chow line. While standing in line there was An E-4 going on watch relief ahead of me. He commented to the E-4 cook serving the meal that "This food sucks!" The cook replied, "Go fuck yourself!" The E-4 in line looked at me, and asked me "Chief, what are you going to do about the cook telling me to go fuck myself?" I instantly recognized this was a situation I didn't need to be involved in. I looked at the E-4 and told him "He didn't tell me to get fucked, he told you. So you can either book the cook (bring him up on charges) or go fuck yourself, your choice." And with that I kept moving on down the serving line. He must have done the later because I never heard anything of it. You're going to come across this a lot. Junior enlisted carrying on and trying to suck you in. If you don't need to get involved stay impartial.

You have to constantly have your head on a swivel. Some of your subordinates will be trying to challenge you and your authority, knowingly or unknowingly. Once again it's just human nature. You control a lot of power over things that are important to them. You control when liberty is granted, leave, marks, awards, discipline, and so on. And if they can control some of that, they will. Basically you're the parent and they're the teenager.

When dealing with subordinates who get into trouble you'll have to learn to stay the ethical course and trust in the system. Because they'll try to derail justice by getting you to drop the charges or look the other way just this once – anything to get themselves out of trouble.

I was a chief on a cutter and one of my E-5's came to me through the chain of command claiming he had been assaulted in the berthing area by an E-5 from another division. Witnesses indicated that the other E-5 was antagonizing my E-5 and in defending himself they got into a fight. After talking to my E-5, I explained to him that I could try and mediate the issue, which I preferred because it kept it at a lower level. Or he could press charges which would take it into the realm of the Uniform Code of Military Justice (UCMJ). Agreeing with me, my E-5 asked if I would try to mediate the issue first. So I arranged a meeting and sat down with the two of them along with the other E-5's chief.

The E-5 from the other division didn't want to take any responsibility for his actions. Even though there were witnesses to attest that he was the aggressor. With the meeting going nowhere my E-5 decided he wanted to press charges. The other E-5 complained that was unfair. I said to him that if he wasn't acting like he was "back on the block," and picking a fight, there wouldn't be an issue and we wouldn't be here.

The aggressive E-5 made the statement that he found my remark offensive and if my E-5 didn't drop his charges he was going to press charges against me for what I said. I told the E-5, "Hey - do what you want, but your issue with me is separate from your issue with my E-5." My E-5 eventually went forward and pressed charges for assault.

At the Captain's Mast to conduct the UCMJ proceedings the commanding officer heard all the details along with witness statements; and found the other E-5 guilty. After being found guilty the aggressive E-5 told the commanding officer that he would like to press charges against me for my insensitive comment to him; referring to him as being "back on the block."

I had no idea what could be offensive about that statement but you never know what someone is going to find offensive. It was common in my generation to refer to where you came from as "back on the block."

The aggressive E-5 didn't want to take responsibility for his actions. When he saw that he was trapped he tried to blackmail his way out of it. And when he was found guilty he didn't want to go down alone, he wanted to take whoever he could down with him – in this case me. Fortunately, the commanding officer had the same feeling about my statement and didn't find it an offensive remark.

So just stay focused on the issue and maintain your moral composure and standards. If you do, it will run its ethical course. It didn't matter if I was in the right or wrong with my statement that the E-5 found offensive; it had nothing to do with the issue of him assaulting another E-5. If the commanding officer felt that I had committed an offense, it would have been dealt with as a separate issue.

In a similar situation dealing with ethics, I was a master chief afloat and one of my collateral duties was the command master chief. There was an issue involving a male E-5, female E-4, and a rated E-3. The rated E-3 was making his rounds while on watch and came across the E-5 and E-4 violating an article of the UCMJ. There were no other witnesses and it came down to a "he said she said" issue. Because the E-5 and E-4 where stating that the rated E-3 mistook what he saw them doing. But since the rated E-3 had claimed and reported that there was a UCMJ infraction, the E-5 and E-4 were brought up on charges and an investigation conducted. The investigating officer found enough facts to recommend the issue proceed to Captain's Mast.

In preparation for the Captain's Mast the E-5 and E-4 came to see me for counsel. Which was normal because that's part of what I did as the command master chief; provide wise council. They felt that if it was their word against the rated E-3, the commanding officer would side with them; whether they were telling the truth or not. I advised them to tell the truth. And if they were guilty to admit it. I said for one, it's the moral and ethical thing to do. Furthermore the commanding officer would more than likely go easier on them if they admitted their guilt. If he caught them in a lie he would throw the book at them.

At Captain's Mast the E-5 and E-4 admitted that the rated E-3 was correct and they were violating an article of the UCMJ. To my disbelief the commanding officer threw the book at them. Thirty days extra duty, 30 days restriction, and reduction in rank.

After the mast, in private, I told the commanding officer I thought he was too harsh. He told me he was setting an example. First, he wanted to let the crew know that violating that article of the UCMJ would not be tolerated. And second, that their initial lying to the charge wouldn't be tolerated. Okay, I found that fair enough. However the E-5 and E-4 were pretty upset with me. They felt that I had steered them into telling the truth and I gave them the impression that if they did they would get off lightly.

Well I wish it would have turned out better for them. And I really thought the commanding officer would go lighter on them. There is nothing worse than seeing two promising careers get derailed, whether they deserved it or not. But there was no way I was going to encourage them to lie.

What kind of ethical example would that be setting? This is a big point to make, never get caught in a lie. And the best way to never get caught in a lie is to always be honest and tell the truth. If they had told the truth from the start the entire issue could have possibly been handled at a lower level. And that's what you should strive to do if you can.

I was a lieutenant junior grade (LTJG), yes you read that correctly, a LTJG. As a master chief I accepted a commission and transferred over to the officer corps as a LTJG. Yeah, that's a whole other book! I was on my first or second patrol and had just taken over the responsibilities of the fueling officer and in charge of fueling evolutions. I had relieved an ensign who had it before me. Before fueling at homeport and various port calls I had to make sure there was funding provided for the fuel. The supply department was responsible for providing me the paperwork. The supply department E-6 was an older grumpy guy. Not able, or wanting, to be recommended for chief he was happy to be a career E-6. I don't have a problem with that, and I'll explain this in the next chapter, if you're not capable of handling the responsibility don't take it on.

Well I started to have a problem with the E-6 getting me the paperwork on time. So I talked with him and let him know my expectations. I got the impression from our conversation that he was used to bullying the naïve ensigns before me into doing things his way.

My conversation with him sort of resolved the issue for a few fueling evolutions, but the paperwork was still not provided to me on time. And I was having to search for him on fueling day. His attitude was that I was bothering him; probably like the ensigns before me. I decided to have a talk with his chief and let him know my issue. The chief also seemed intimidated by the E-6 and I got the impression that the chief just let the E-6 do what he wanted.

I could see the writing on the wall and that this was going to come to a head sooner or later, and began to document his behavior. Like most people I really don't like confrontation. If it comes, I like to think that I have no problem dealing with it. But it's not something I enjoy. I was always one to give people the benefit of the doubt and kept giving this E-6 chance after chance.

A few days later we pulled into a foreign port, and the following day I set up to take on fuel. Before the fuel supplier would start to transfer fuel they wanted to be paid, so I went looking for the E-6 to see where my paperwork was. I found him hung over in his rack. I told him, "Get out of the rack and get my paperwork to me on the pier so I can commence fueling. Now! That's an order!" About 5 minutes later the E-6 came out onto the main deck, about 20 feet above where I was standing on the pier. But instead of coming down with the paperwork, he stood at the railing and yelled at me about waking him up for the paperwork.

This was happening in front of the refueling detail which is made up of junior enlisted. Now they were interested in seeing how the old E-6 was going to put the junior officer in his place.

I calmly directed him in a firm voice, "Get the paperwork now and bring it to me here on the pier. That's an order." I think he could tell by my calmness that I wasn't putting up with him today and he complied. The next day I had him, his chief, and the command chief sit down with me. I told the command chief we could do it in the wardroom or chiefs mess. He picked the mess. I was sitting along with the chiefs and the E-6 was standing. I addressed the E-6. I told him that I had been documenting my issues with his performance over the past few weeks. And that I was filing UCMJ charges against him. I could tell by the look in his eyes that I had his undivided attention.

I then turned to the chief and told him that he was now responsible for getting me my fueling paperwork. And if I didn't get it on time I would start documenting his performance.

The E-6 asked me if I could give him another chance, that he was sorry for his actions. He said he took full responsibility for the late paperwork and it wouldn't happen again. I said "Haven't I given you enough chances already?" He replied that I had and he was sorry he hadn't acted on them.

I thought about if for a minute and told him that I would put him on probation. If I had one more issue with him I would file the charges. I had every right to file them and he would've more than likely been found guilty. And he knew it. But this was one of those situations where I could use the power of the UCMJ without actually bringing the E-6 up on charges. I would like to think that it was an overdue wake up call for him to get his act together. And it worked, because I didn't have a problem with him the remaining year and a half we were on board together.

Why give the E-6 one more chance? Because I feel that your last resort should be UCMJ charges. And now he knew without a shadow of a doubt that he controlled his own destiny. If you ever have the option of placing the success or failure of an individual in their own hands, do it. Then they only have themselves to blame.

If the E-6 had a stronger chief over him I believe he would've had a different attitude, or he would not have been in the service anymore. Because the chief would have given the same ultimatum that I had the first time the E-6 blew off his responsibilities and superiors.

But unfortunately the chief wasn't a good leader. In fact the chiefs' mess on this cutter in my opinion was weak.

And I'll explain that further in the chapter "Setting the example."

This is probably a good a time as any to bring up placing someone on report, or really bringing any type of disciplinary action against someone. Bringing someone up on charges, or booking them, is really the last thing you want to do in the way of discipline. But sometimes it's just cut and dry and you have your hands tied. The individual has left you no choice. I had an E-6 caught with a loaded handgun in his locker during a berthing area inspection. And if you think that may sound odd, it happened on two cutters I served on. Of course the inspection was being conducted by the executive officer and the member's chief was directed to place the E-6 on report.

Then it gets a little murky. Like, if the E-3 in your watch section is 20 minutes late for watch relief. Do you blow it off, council them, write a negative administrative report, or bring them up on charges? Personally, if it was their first offense I would council them. If their second, negative administrative report. And so on.

We can really get into the weeds here with all the different scenarios you can run into. The bottom line is that if a rule or regulation is broken, and you know about it, you're going to have to make a decision; do I act on it or not? If you *do* act you're going to need to decide what degree of punishment. If you *don't* act, you may have to answer to your supervisor.

Let's take the previous example of the E-3 being late for watch. If you don't do anything, that's your call. But let's say an E-5 in another watch section gets upset that the E-3 got a pass. Because a few weeks ago he received a negative administrative report from another supervisor for the same thing. He's so upset that he brings his grievance up to the executive officer during request and complaint mast. Now the executive officer calls you in to explain yourself.

So when you take a position on an issue make sure you're ready to back it up. If you're not ready to back it up, you're not ready to make the decision. And when you get it wrong, like I did with letting the E-3 go home without leave, you have to own up to what you did.

Finally, let's talk about your positional power going to your head and getting you into trouble. It happens more often than it should. Here's a random example I pulled from a recent Coast Guard Judge Advocate General disciplinary report. I say random because unfortunately I had plenty of convictions to choose from;

"An E-7 was convicted by a summary court-martial of maltreating three enlisted members subject to orders by relaying harassing/indecent language, showing videos of a sexual nature and taking unsolicited photographs and sending those photographs to another member,

assault committed against three members; by striking one on the face, pressing down on another's abdomen with their palm, poking two of the members in the ribs, striking one on the thigh, pulling on the shirt sleeve of one, hitting a member on the head with food, and poking a member on the back. (Violations of Articles 93 and 128, UCMJ). The member was sentenced to a reduction to pay grade E-6."[6]

Do not think because you're a chief that you're above the law! The most common over reach of positional power I've seen is inappropriate sexual conduct. And the second is covering up the inappropriate conduct. And by the looks of recent Coast Guard disciplinary reports this is still the case. So be forewarned that you could end your career by abusing your positional power. And along with ending your career, you tarnish the anchor you wear; both needless tragedies.

Chiefs Mess

Dealing with your fellow chief is probably going to be one of the toughest parts of your job. To the new chief this statement may sound odd. But to the seasoned chief it's completely understood. Because your fellow chiefs are just like you; opinionated, bold, outspoken, fights for their people, and thinks they know it all.

So if your impression of the chiefs' mess is a place where you can catch up on your movies and have a mess cook wait on you – you're mistaken. The chiefs' mess, for the most part, is where issues and policy are hammered out, debated, and fought over.

In a dysfunctional mess there is no leadership. Chiefs avoid each other and they avoid discussing issues. Not only is there no esprit de corps but there's a mixed message coming from the chiefs' mess. Because each chief is pushing their own agenda.

A good chiefs mess has a leader to guide them, mutual respect, discussion - sometimes heated, and opinions expressed and challenged. But when a decision is made they show a united front. And most of all -chiefs that hold each other accountable.

If all that sounds a bit much for you maybe you're not ready to be a chief. I know that is a strong statement but it's one you have to ask yourself. Because it's not for everyone.

Before 1995 there was no high year of tenure in the Coast Guard. You could hang out pretty much as long as you wanted in any paygrade until the age of 62, without the fear of being discharged for not promoting. Most of the E-4's through E-6's retired at their 20 year mark and the E-7's through E-9's hung around longer. If you were incapable of holding a paygrade you were demoted or discharged. In my opinion this worked out pretty well. Because people found where they were comfortable in the rank structure and no one pushed them to promote into leadership positions they didn't want. Not all of us are cut out to be leaders, responsible for other people.

Then along came high year of tenure, a force management tool. To start with, the Coast Guard only enforced a 30 year cap on active service. Later, they slowly phased in time in service and time in grade requirements for all ranks. This caused people to get out of their comfort zones, and into leadership positions by forcing them to advance. And in my opinion this forces people into leadership positions they're not ready or willing to handle, thus setting them up for failure.

This is how you get people promoting to chief, not because they are ready for a senior leadership position, but because they want to stay in the service. So, who should step up and say that a chief isn't ready or capable of handling the rank? Well, if the chief's chain of command - division officer and department head - don't step in I feel it falls on the chiefs' mess to police their own ranks. If in the opinion of the mess, a chief is found unfit to be holding that rank, the command chief should consult with the commanding officer to have that chief reduced in rank or processed for discharge.

Unfortunately if no one intervenes, the chief in question will eventually screw up and cause their own demise by UCMJ action or discharge. But by then they may have more than likely demoralized their subordinates, ruined careers, and caused others to vacate the service in disgust.

Good chiefs need to be proactive in the mess and challenge their fellow chiefs. Interacting with your fellow chiefs and being in a passionate chiefs mess is a challenge. And you're going to bump heads. The key is to not take it personally and to not let it get out of hand.

As a newly initiated chief I walked across the brow of my first assignment relying heavily on all the power and prestige I thought the position granted me. Because, truth be known, I was nervous about my new responsibilities. Ultimately, this helped me in a situation with another chief, because it gave me confidence. Too bad that didn't stop the situation from getting out of hand.

It happened with a fellow engineering chief. We worked in the same department but were in charge of different divisions. He was just as stubborn as me but had a hair trigger temper. He was a great technician and he and I worked closely because we were both in the engineering department. But he had been a chief longer than me and felt he could get over on me with his seniority. And this just didn't sit right with me. Especially when he argued technical responsibilities with me in front of our subordinates. Because if I lost, my guys lost by way of respect for me and extra work.

And I didn't appreciate him putting me in that position. His problem was that he wanted to be the lead chief instead of being equals, the way it should've been. Not only for our own working relationship but that of our subordinates.

After we had bumped heads a few times I talked with the command chief about the situation. He talked with the other chief but it didn't resolve the situation. And then I went to my immediate supervisor, a warrant officer. He just wanted me to back down and have the situation go away. But that just made me dig in my heels. I felt that if I backed down, it would be detrimental to my guys because the other chief would have the upper hand in determining what work his division would do; aka we got the crappy jobs.

It all came to a head one evening while we were underway and I was standing engineer of the watch in the engine room. The other chief was upset about me working on a piece of *his* equipment which was located in *my* engine room. He came storming down into the control booth where I was standing watch and started ripping me about getting into his equipment. As usual he didn't care who was around so I cleared the control booth of junior enlisted and we had words, and I had enough. I called the engineering officer on the phone and asked him to come down because things were about to get out of control. By the time he got there it was too late. The other chief and I had a death grip on each other, cussing, and trying to choke each other to death. Yep, it was embarrassing! The engineering officer was a great supervisor and I felt like an idiot letting him see me stoop to school yard shenanigans. But it was one of those situations where I had to stand my ground. The other chief called me out in front of my subordinates, putting me in a no win situation. A better chief would have done it in the chiefs' mess in private. We could have beat the crap out of each other in there and it would have been between us.

But the end result was that he didn't bother me or my subordinates after that. And we gave each other a wide berth from then on until he eventually transferred. And his replacement and I got along just fine.

No doubt about it, you're going to bump heads with a fellow chief at some point. And I hope when you do it isn't as bad as what I just mentioned.

But like I said earlier; we aren't all perfect. Obviously there were some respect and personality issues working against us. It happens. You're not always going to get along with the people you work with. And if you have issues with a peer you need to resolve them in the mess.

I had taken over as the command chief for a patrol while the regular command chief stayed back for some medical issues. During the patrol one of the enlisted crew members got into trouble and was being taken to Captain's Mast. The issue was a hot potato in the chiefs' mess. Half of the chiefs thought the matter should be handled by the chiefs' mess while the other half thought the member should go to Mast. Handling issues at the lowest level is always your first option. And letting the chiefs mess handle a discipline issue would save any permanent entries in the offender's service record. In handling the offender the chiefs would council them and administer light extra duty. And the offender knew if they screwed up again they were headed to Mast.

The executive officer was compiling all the statements and findings for the Mast, and had asked my opinion as the command chief. He said that there was enough evidence to take the member to Mast. And that in the command's opinion the infraction required a UCMJ proceeding. The chiefs' mess was divided on this. Given the weight of the infraction and the mess split – I decided the member should go to Mast.

The executive officer suggested it would be a good idea to have the chiefs mess present at the Mast to show support for the command's decision, whatever that might be. I agreed and asked the other 7 chiefs to attend the Mast. Normally just the command chief was present and the accused member's chief made an appearance as a character witness. The Mast was held in the wardroom and was open to the crew. This was typical for this command's Captain's Masts so the event and process would be seen as transparent, and to show the crew what getting into trouble looked like.

The enlisted member was found guilty and punishment was handed down by the commanding officer, which didn't involve the chiefs. After the Mast we assembled in the chiefs' mess for lunch. One of the more outspoken chiefs in the mess was sitting in a chair across the table from me and he was fuming. He didn't like the fact that I had made the final call to send the crewmember to Mast, and felt that the chiefs had been manipulated in attending the Mast. He assumed that by being there we would have input on the level of punishment. He was yelling at me and accusing me of lying to him. Words were exchanged, and he stood up and threw his chair at me. He literally stood up and threw his metal chair at me from about eight feet away.

I don't know how he missed me! The chief was a good guy and we had our spat in the mess, but when we walked out of the Mess we put it behind us; we were chiefs and acted as such in front of the crew.

And that's the way my first story should have ended, with us having it out in the chiefs mess, not in front of the crew.

And so there you have two stories, although perhaps on the extreme side, they demonstrate the passion and volatility of a chiefs' mess. I could go on and on with stories about the passion of chiefs arguing in the mess, but you probably already have some of your own and get the point.

Let's get into holding each other accountable. It's not easy to tell another chief to square themselves away. Most, like the chiefs in the previous scenarios, are going to have contentious personalities. So you have to do it tactfully. And this is something you'll run into more than you think. Because there are chiefs who think that they're above the same rules and regulations everyone else is required to adhere to. I'm not talking about major infractions, just the little bending of a rule. And they usually justify their action by telling themselves that rank has its privilege. So they push, or just cross over the limits of a regulation here and there. I know because I have fallen victim to this way of thinking myself.

I was a master chief on independent duty and occasionally worked with another master chief in another command. He was the model for the paygrade. He didn't cut corners. His appearance and uniform were always squared away, complete with shined boots.

He just *looked* like a master chief, with a high and tight haircut and a chiseled face. When he was walking toward you, you knew he was somebody important just by the way he looked and carried himself. He should be writing this book instead of me. But God bless him he passed away much earlier than he should have.

Well, while shopping at the exchange one day I saw this really neat master chief belt buckle. It was brass and had the master chief emblem of a Coast Guard shield, anchor, and two stars on it. I bought it and started to wear it with my uniform. Only a plain brass belt buckle was in regulation and this one with the emblem was intended to be worn with civilian clothes. I didn't care, who was going to tell me to take it off? All the places I normally went to at the time had, at most, a senior chief or very rarely a junior officer around. They weren't going to tell me to take it off.

One day I attended a meeting with a group of junior enlisted, and my friend the squared away master chief was there. He noticed my belt buckle during the meeting, but being the professional that he was he didn't say anything until we were alone later that evening. We were going over the day's events and he casually said, "Ed you know that belt buckle you're wearing is not in regulation." I said, "I know." That was it, we didn't say anything else. Later that evening I thought about what he had said and the way he had said it. I didn't wear that buckle in uniform again. I had that much respect for him. But I'm sure if I had worn it again, my fellow master chief would have called me out on it again; although probably not as nicely as before.

I'm not going to try and kid you and say that every situation will be that easy and the guilty individual will comply. But if your approach and tone is right you'll be surprised what will happen. And if not, then kick it up a notch. But unless you're dealing with a major infraction you won't need to pull out your sledgehammer to tack a nail.

Another part of a chiefs' mess that I feel strongly about is community involvement. Get involved with other chiefs and leaders in your community. I'm talking about the fire chief, police chief, mayor, local politicians, and so on. People you may have to deal with at some point. I picked up a really neat way of doing this from a warrant officer we used as a chiefs initiation judge.

What we would do is have the prospective chiefs going through initiation pick one of the previously mentioned officials and set up a meeting with them. We would instruct the prospective chief to present themselves, explain the mission of their Coast Guard unit, and get a picture of them with the official. Because we had to have proof that they met! And it was a nice keepsake for the prospective chief.

This worked out great, and through the prospective chiefs we started to network with other agencies and officials in our area. Trust me, it helps if you have already introduced yourself to the chief of police before you receive a call from him that he has one of your guys locked up.

Another bridge building endeavor is to get to know the other services near you. If you're at a land unit see what facilities are around. All of them have senior enlisted organizations. The Navy has their chiefs' mess and the Air Force their Top Three Association; to name a few. Get to know these folks. Networking with these people can be very beneficial. If you're afloat and pull into a port call go over and check in on the vessels moored up with you, even if they are foreign flagged.

When I was the command master chief on an afloat unit and pulled into a Navy base, the base command master chief always checked in on me to see if there was anything the mess or crew needed. I really appreciated that gesture.

On one port call I had several prospective chiefs preparing for initiation. We pulled into a Navy base and the command master chief came aboard as usual. I told him that I had some chiefs getting ready for initiation and asked if I could bring them over to his chief's mess. He obliged and put together a quick party for us. We had a great time.

And I'm sure our prospective chiefs will never forget engaging with those Navy chiefs!

I was also able to return the favor to a Navy submarine chiefs' mess. They had pulled in across the pier from us after a long time between port calls. I went over and introduced myself to the sub's command chief and offered our chiefs mess as a place to relax. And if you have ever served on, or toured a submarine you'll understand that they really appreciated having the extra room to stretch out.

And there is always your local Chief Petty Officers Association (CPOA). Most major Coast Guard units have a chapter near them. The CPOA is a great way to get in touch with the vast "storehouse of knowledge" of retired chiefs. Not only will you benefit from the experience but they would enjoy interacting with active duty chiefs.

I guess this would be a good time to address the chiefs' initiation. Although voluntary I would recommend all chiefs take part in the process. And if you're an initiated chief you should support and attend the initiations. What is misunderstood by some, and thought of as hazing, is actually a well-planned team building event. Team building with your peers and superiors.

Trust me when I say that the only thing that may get hurt during the initiation is your feelings.

But that's a major part of the initiation; teaching you to be humble. And humility is a reoccurring theme in the initiation, the Chiefs Creed, and even this book.

As I said the initiation is voluntary. Nothing states that you have to go through it. And what will happen if you don't go through it? Well, you'll just miss out on the tradition of being welcomed into the fraternity of chiefs. Will you be shunned if you decline? Over my career I have known several chiefs who declined to go through initiation. They took the usual jeering for a while and then the issue went away.

The initiation is a big deal in any chiefs' mess. At some point the uninitiated realize it is tough to be excluded from such an important event. Ultimately most decide to go through it after they have settled into being a chief and realize what it's all about.

I will have to say that the initiation has rightfully been modified over the years to make it a more meaningful event. It can't be just a humiliating experience or its nothing more than hazing; which is illegal. So it's up to the chiefs' mess to continue to monitor and modify it when needed, and most of all make sure it is conducted within the Commandants or Chief of Naval Operations guidelines.

Finally let's talk about rules in the chiefs' mess. Every mess I have been in has had one rule in common; *respect your fellow chief.* Any other rules are at the discretion of the chief of the mess.

Unlike the wardroom I have never encountered any seating order at the meal table. I have been in messes where the mess cook set the table and we all ate at the same time - family style; watch reliefs excluded. I like this arrangement and encourage it for at least the lunch meal. Because it sets aside one time a day when the mess can get together. No agenda should to be established, just casual conversation.

Some required the payment of dues to cover incidentals within the mess; like tipping the mess cook, stocking the pantry, and so on. So you'll just have to check with your chief of the mess.

If you're in a mess with fellow chiefs (E-7's) you're obviously on a first name basis. But if you have senior and master chiefs you'll want to address them as such and respect their rank. It's their call if they allow you to address them by their first name in the mess.

Watch what is said and done around the mess cook. Whatever is said or done in front of them will likely be relayed to the entire crew.

Command Chief

I decided to make my thoughts on the position of command chief a separate chapter instead of including it in the "Chiefs' Mess" chapter because of its complexity. And I'm including it in this book for newer chiefs because you may find yourself as a command chief on your first assignment as a chief. I wasn't assigned as the command chief at my first unit, but did fill in several times during that tour. So it might happen.

In the Coast Guard the command chief is usually a collateral duty at units below the district level. And that's what I'm talking about here; units below the district level that do not have full time trained command chiefs. The units where you're going to cut your teeth as a chief.

Your unit's commanding officer will always assign a senior enlisted member as their command chief. The position is filled by the commanding officer, serves at the pleasure of the commanding officer, and is the commanding officers liaison with the crew.

Within the afloat chiefs mess you may also hear the terms chief of the boat (COB) and chief of the mess. Usually these two positions along with the command chief are held by the same person and considered one; but not always. The COB is the enlisted person with the highest seniority in rank. And this is usually the person selected by the commanding officer to be their command chief. But if the commanding officer doesn't have faith in them, or the COB declines the position, you can have two separate people as the COB and command chief. And if you have two separate people in these positions, you can have the COB running the chiefs mess (chief of the mess) and the command chief running everything outside the mess.

As a general rule the commanding officer doesn't care who is running the chiefs mess. They just want it functioning properly. But they are certainly concerned with who their command chief is, because that's who they're going to be dealing with as the representative for the enlisted crew.

Like I mentioned earlier; you may find yourself as the command chief sooner than you think. Because those senior can simply pass on the job. You'll run into chiefs that don't want the extra responsibility of the command chief. Because it's a major collateral duty that eats up a lot of their time and distracts from their primary billeted specialty. And there's also the fact that the commanding officer may not have faith in them.

At the time of the chair throwing incident, I was one of the more junior chiefs in a mess of eight chiefs. But nobody wanted to step up and take on the command chief position when the incumbent had to stay back for a medical issue. So you may find yourself doing the job sooner than you think. Or if you're assigned to a unit with just one chief; you're automatically it.

So what are your responsibilities as a command chief?

Foremost you're responsible for the care and well-being of the enlisted crew. So you should be monitoring the pulse, morale, attitude, or whatever you want to call it, of these people under your care.

You should be one of the first people they are required to see when reporting aboard. You want to impress on them the importance of using their chain of command, but they should know that you're approachable and available 24/7.

You'll need to process issues and concerns. And if needed, address them with the commanding officer. Although most of your interaction will be with the executive officer. But know that you have a direct line to the commanding officer if needed.

Handling low level infractions/complaints. Although you should be approachable all the time, I also found it helpful to schedule a weekly time frame for crew members to come see me. You should do this one-on-

one with the individual. See what's going on first before you have their chief present – because their chief may be the problem. When I was dealing with a member of the opposite sex I asked the member if they would like another female present. Talking to a person of the opposite sex can be difficult in some circumstances and having another female may help them to open up with their issue.

From here handle the issue at the lowest level possible. A majority of the problems you will be dealing with are just plain gripes. Like "I stood watch last Christmas", "They always want to watch the same movie in our lounge", and so on. These are important enough to the individual to bring them to you so you have to at least entertain them. And you'll have members coming to you because they *don't* want their chief knowing what's going on with them. And you'll have to judge if this is valid or not. If it is valid, respect their privacy and don't turn around and discuss the issue with their chief later in the mess. If it's not valid explain to them the reason their chief needs to be involved. And in my experience you want to get their chief involved at the earliest part of the problem. Because whatever is going on with the individual is going to impact some aspect of their work or interaction with co-workers, especially in their division.

When disciplinary action needs to be taken on an individual you're responsible for making sure the member is given due process. If possible you should try and negotiate with the executive officer to have infractions handled without going to Captain's Mast. You should be present at every Captain's Mast involving an enlisted person to make sure they are properly represented by their Mast representative.

I found that most commanding officers go into a Mast with an open mind. They stay out of the pre-Mast process and make their determination upon presentation of all the evidence and testimony at the Mast. That is why you're dealing with the executive officer on pre-Mast issues. I always made it a point to engage the commanding officer in private after the Mast to see what influenced them in making their decision. This helped me with counseling members preparing for subsequent Masts. And it gave me the opportunity to voice my opinion, ever so respectfully, on the judgement.

Be a liaison for the ombudsman. Usually they have a direct line to the commanding officer, but you should request to be present when they meet. This will help facilitate any issues they may have. Be proactive with your ombudsman and with the ombudsman program. When families feel involved with the unit, it tends to reduce the number of crew members coming to see you with problems. Because many issues you'll have to deal with come from members' personal lives and not their professional lives.

Promote enlisted recognition programs. You need to be proactive with enlisted recognition programs such as sailor of the quarter, enlisted person of the year, and Douglas Munro award. You should promote these at your unit and submit qualified candidates for district, area, and Coast Guard wide levels. Keeping the crew engaged in their pride and professionalism is a win-win for everyone.

Don't be afraid to go outside of your unit's chain of command. And if you feel that you have an issue that is not being addressed by your commanding officer don't forget that you have a second chain of command outside of your unit. You should address your concerns with your district or area command master chief. Never forget that you have this option to circumvent your unit's chain of command.

You and your junior officer

Next to the seaman recruit reporting aboard straight out of boot camp, there is no more highly motivated person crossing the quarterdeck for the first time than the newly minted junior officer. And like the new seaman recruit, the junior officer needs to be trained. And this is the reason you were paired up with a junior officer when you went through chiefs' initiation. It's your job to help mentor them.

And here is why they need mentoring; the academies spit out hundreds of junior officers a year who have gone through four years of the best education and leadership training money can buy. So they are smart people. But that doesn't mean they have been prepared for the scenarios they will encounter in the fleet.

Cadets live a regimented life within the bubble of the academy. Their only exposure to the real military is maybe a summer cruise they will take in their junior or senior year. They aren't taught how to interact with senior officers twice their age in the wardroom. Nor are they taught to interact with their division chief.

And humility is definitely not part of their curriculum – precisely the opposite. To be blunt, your junior officer will most likely come to you with a chip on their shoulder. Your job is not to smack it off. It is to nicely take it from their shoulder and place it in their hand.

And this won't be hard to do. Because when they report aboard, and I'll use the example of a cutter, they are loaded down with a very heavy schedule. So here is an eager junior officer who has a division to run, damage control and watch qualifications to complete, plus at least a few collateral duties. They're like the medical intern at a hospital; tired, over worked, and totally stressed.

But most junior officers aren't going to just come out and ask you for help because they don't have the humility to ask you. In fact they're probably going to be intimidated by you. You're more than likely twice their age and you're the chief. Remember that this first assignment is basically a training billet for them. It's to get them acclimated to the real world of being an officer. This is why you need to sit down with them and, like at your initiation, explain that you will work with them to help them be successful.

Of course there is always the one junior officer who thinks they're an admiral and you're not going to tell them anything. Just give them a wide berth and in due time, when they get bogged down in all their qualifications and duties, they'll come around.

Don't be the chief that tells them to stay out of the way and you'll handle things. You're not doing them or your division any favors.

I was an E-4 in the engineering department. On board was an ass chewing chief in deck department that ran everything in his division and probably most of his department. The chief was a tall, older guy who just naturally looked intimidating. Everyone on board gave him a wide berth because his ass chewing had no departmental boundaries - everyone was fair game, including any officer he could get over on.

One day our new auxiliary division officer, an ensign, asked me to get the rigid hull inflatable (RHI) small boat ready. He wanted to take it for a spin to make sure recent repairs made to the engine were correct. A little out of the ordinary for him because our chief should have been handling technical issues within the division. But this ensign was fresh out of the academy and full of energy. The RHI belonged to deck department and we just maintained the engine. I didn't think anything of it. So we loaded up and started to spin around the pier area there in our homeport. After about 30 minutes I looked over toward the pier and there was the chief of deck department, smoking a cigarette and motioning with his finger for us to come over to him. I knew that this wasn't good. As we maneuvered the RHI over to him I could tell that the young ensign, who out ranked the chief, didn't know what he was about to get himself into.

Now in the military we have what are called positional authority and coercive authority, and the young ensign was about to get a lesson in the difference. When we got close to the pier I could see the chief's face twitching. I knew that twitch and wished I was anyplace else. The chief yelled at the ensign "Why in the fuck are you driving around in my boat without me fucking knowing about it?" There was no place to hide in the tiny RHI and I just kept my mouth shut. After the ensign got a good ass chewing we took the boat back to the cutter in silence.

Of course this was back in the day before high year of tenure and you could stay in until you were 62. Most of the chiefs back then were in their 50's and 60's and had no interest in mentoring a 22 year old junior officer. But this still goes on today. Sure you can get away with pulling coercive authority over on a young naive officer. But what good does that do? In this case the chief's own junior officer was locked up somewhere afraid to get near their chief let alone question him on anything. Obviously this chief had no respect for my chief or he would have gone to him about us being in his RHI. And after he tore into my new division officer, who never spoke a word in his own defense, I immediately lost all respect and confidence in him. And in the end that was one of my worse assignments.

Because the levels of power and authority with the officers and chiefs was all ass backwards. No one knew their place in the organization, or at least wanted to stay in their place. And it was on full display for the junior enlisted crewmembers.

Here's a better, although not perfect, way to handle a situation with a junior officer:

I was a master chief and command master chief, so I pulled a lot of weight onboard the cutter. I was standing in the chow line and had let the junior enlisted watch reliefs ahead of me, per the captain's standing order. Chiefs and officers had head of the line privileges behind watch reliefs. Well here comes one of the new ensigns who had just reported aboard and goes straight to the head of the line. The junior enlisted knew she wasn't a watch relief and started grumbling.

This is another one of those situations where a decision needed to be made, whether to get involved or not. Because I also knew she wasn't a watch relief. Seeing that there was a standing order violation taking place, and I was the senior enlisted present, I decided I needed to say something. So I asked her, "Excuse me ma'am but are you a watch relief?" The answer was "no." I said "I know you're new on board ma'am but these enlisted watch reliefs go to the head of the line." The ensign stated "Thank you master chief but I know what I'm doing." I then said "Very good ma'am, we can talk about this later." I felt that at least I had brought the issue up in front of the enlisted crew so they knew I wasn't looking the other way and condoning her behavior.

Now I could have gotten into it with this ensign, and really wanted to. But that would have just replayed the whole previous story over again. Or the ensign could have pulled rank on me and shut me down. You never know what's going to happen when you call someone out in public.

We had great leadership onboard and I wasn't about to mess it up by tangling with this ensign in the enlisted chow line. So after chow I caught up with the ensign in the wardroom. I respectfully asked her why she was getting ahead of the watch reliefs if she wasn't one. She told me basically that rank has its privileges. I professionally advised her that rank has its privileges but those don't include breaking the rules. She understood what I was getting at. I excused myself and went about my business.

So don't put yourself or your junior officer in a bad position publicly. Someone will come out the loser; you, the junior officer, or the junior enlisted witnessing it.

And on that note you want to make sure all the junior officers at your unit are treated respectfully. The wardroom in my opinion has a strange way of treating junior officers, a sort of trial by fire during their first assignment. If they can't handle the authority and pressures as a junior officer they won't make it as an officer. Well that's the wardroom's business. But it's your business to make sure you don't condone the disrespect of authority.

Hopefully, the junior officer will handle any respect issues on their own, but if they don't, you can show them how it's done. A case in point is when you're walking down the pier with a junior officer and an enlisted person passes by without saluting. Of course the enlisted person usually makes it seem like they are engaged in something else or looking the other way. Or they could just be plain disrespectful because they have no respect for that particular junior officer. However, the enlisted person is not only disrespecting the junior officer, but you as well. If the junior officer doesn't correct the situation you should stop the enlisted person, engage them, and have them render the proper courtesy to the officer. In one action you will have modeled good discipline for both the officer and the enlisted.

You'll also want to keep an eye on your junior officer as they interact with your subordinates and other junior enlisted. Your junior officer is not going to be your age, nor that of the rest of the chiefs and probably a majority of the wardroom. They're going to be the same age as your E-3, E-4 and E-5's. And that's who they're going to naturally gravitate toward. The next thing you know they're running into each other on liberty and hanging out with each other. Then one day when they're walking down the pier, the officer let's one of their enlisted buddies get away with not saluting them. After all, they don't want to be considered a hard ass. If you notice this, mentor them about how inappropriate this is. Refocus them on their responsibilities as an officer.

Something I found that junior officers are really good at is writing. They can be of great help to you in that area. Even though I had lots of experience, I still had trouble writing marks, administrative reports, and awards to suit my supervisors. You should tap into your junior officer's expertise. They live in a world of officer evaluation reports (OER's). And if you've ever read an OER you know that there is a lot of "wordsmithing" and preparation that goes into them. So after you have a cut on whatever you're producing for your subordinates, ask your junior officer to review it. This will save you a lot of frustration and re-writes.

Don't forget the Coast Guard is small, and that junior officer just might be a commander or captain down the road. No matter what career path you take; staying enlisted or taking a commission, you're never going to catch them in rank. So do yourself and the Coast Guard a favor and mentor them the right way.

You'll be glad you did should they turn out to be your commanding officer one day.

Setting the example

"Chiefs today aren't like they use to be." Yeah we've all heard it. It's like when I first came in everyone talked about how great things were back in the "old guard!" And as long as someone had one more day in the Coast Guard than me they were part of the "old guard." And I'm sure they have a similar saying in the Navy.

And back in this "old guard" was the corps of chiefs who ran everything and carried the Coast Guard on their shoulders, to the admiration of everyone - unlike today's chiefs.

Is this mythical "old guard" standard too high? Well I think it's a standard you should be aiming for. And as a chief your actions and the way you conduct yourself should be admired.

Your subordinates should look at you as the example. When they ask themselves what their uniform should look like, yours should come to mind. When they wonder how they should treat their fellow shipmate, you should come to mind, and so on. Your conduct, leadership, and appearance is on display 24/7.

It's tough. But no one said being the chief was going to be easy.

Get out and hit the deck plates and don't manage your people from the chiefs' mess or your office. Know what's going on. A lot of potential problems can be resolved by having one on one discussions with your people as they work. This is also a great way to give them a sense of inclusion, to mentor them, and to show you care about them.

A standard I found very beneficial was to not direct your people to do anything you're not willing to do yourself. It seems pretty simple but just think about it for a minute. How many supervisors have you seen doing a menial task? Not too many I'd bet. Why, because they think they're above it. Sure there is nothing to say they have to. But occasionally working jobs with your enlisted goes a long way in demonstrating that, although you're the boss, you know what they're up against. Meaning, you know their working conditions, and their interpersonal relationships which can be useful to you when it comes time to mediate issues. And no job is beneath any paygrade.

When I was the chief in charge of the engine room one of the nastiest jobs we had to do was keep the bilges clean. And I expected my guys to not only keep them pumped out, which was the standard, I wanted them wiped clean. Since this was a nasty job it always fell to the lowest ranking sailor to do it. One day I overheard them complaining about having to clean under the locomotive sized main diesel engines. To do this they needed to crawl under them. Definitely not for the claustrophobic! The bilge under these engines was dark, dirty, and in some places only 15 inches high. They said getting under there couldn't be done. I slipped on my coveralls and slid under the #1 main engine and wiped up the bilge, all the way under it from the front to the back. That ended the issue of wiping down the bilges. Not so much because I had shown them it could be done. But because I had shown them that the job wasn't beneath me, and therefore not beneath them.

Something to watch out for is not asking your people to do anything they *shouldn't* be doing. You want to mentally place yourself in their shoes. An example could be something as simple as directing the nearest person to you, say the E-4, to go tell the E-6 to do morning berthing area cleanups. You're in a hurry and wouldn't think twice about directing the E-6. But now you have an E-4 doing it. You have put the E-4 in a vulnerable situation.

I was an E-4 serving in the auxiliary division and one day the chief of main propulsion division was up working in our auxiliary division shop. He was using the huge industrial metal lathe we maintained. There is a shop rule that when you use the lathe you clean the machine so it's ready for the next person.

When the main propulsion chief was finished he left without cleaning up the lathe. I'm sure he was busy with his project and just forgot; he was a chief, no problem. Well my chief comes in and sees that the lathe is a mess and asked me who was using it last. I said, "The main propulsion chief." My chief was pretty upset at him and instructed me to go tell him to clean it up. Now I was only an E-4 and could smell danger all over this. I questioned my chief about me going to tell him, and he said, "Yes you. Now get going!" I could tell he was preoccupied and probably not thinking through what he was asking me to do.

Against my better judgment, but not wanting to disobey my chief, I caught up with the main propulsion chief. He was up forward by 21-man engineering berthing in the main passageway where the ladder heads up to officer country or down to engineering nine-man berthing. The location will forever be burned into my memory. I casually said, "Excuse me chief, but my chief said you need to go clean up your mess on the lathe." Well that's about the time I felt the heat of the main propulsion chief's breath and the wetness of his spit as he was about a millimeter from my face yelling and screaming about how I will never, ever tell him anything! It sounds like a simple statement, but it drug on for quite a while with a lot of profanity and a lot of "Yes, chief!" coming out of my mouth. Needless to say the main propulsion chief didn't go back to clean up the lathe, I did!

So can you see how something you're really not thinking about can turn out really bad for one of your subordinates?

And here's that story I promised you back in the chapter "Know yourself." It really drives home how bad a chiefs' mess can get when no one steps up and sets the example.

I was, as I mentioned a LTJG, serving as an engineering officer in training. As an officer I stayed in my lane and my daily activities revolved around the wardroom. I was no longer a chief and left the chiefs' mess to themselves as it should be. If I had a reason to go to the chiefs' mess I offered the proper courtesy and knocked before entering.

It wasn't long after I had reported aboard that I had the run in with the supply E-6 and his chief. I could tell by the interactions I was having with the chiefs' mess that there were more weak chiefs than strong chiefs.

Well I had some business to take care of with one of the chiefs and went looking for him in the mess. I knocked on the door and entered. To my surprise, one of the female ensigns was reclining back eating popcorn and watching a movie with one of the chiefs. I found this extremely odd. In fact I had never seen that before.

And then while standing watch I overheard scuttlebutt from the junior enlisted that one of the female junior officers, an ensign, was spending a lot of time in the chiefs' mess. She would have her meals there, watch movies, and basically had the privileges of a chief. Their gossip enforced what I had seen.

I tried to help the mess out by recommending to the ensign that she stay out of the chiefs' mess and mingle with her peers in the wardroom instead. I told her she was giving the appearance of fraternizing with the enlisted chiefs. We had already bumped heads on other issues so this didn't go over well, and she told me to mind my own business. That was enough for me. The only rule the ensign was breaking was one that could only be enforced by the chiefs' mess.

While standing watches with the three engineering chiefs I asked what was going on with the ensign hanging out in the mess. They were rightfully embarrassed about the situation and said they had complained to the command chief about it. But it was a very well-liked chief with a strong personality who was inviting the female ensign into the mess. To top it off, the chief causing the problem was also well liked by the commanding officer. So the mess felt like their hands were tied over the issue, as not even the command chief wanted to intervene.

Now, the command chief was a nice guy but I got the impression he was more interested in his rated job than running the chiefs' mess and enlisted issues. He had the position because of his seniority. And probably should have passed the responsibility on to another chief who had more interest in the job.

It's no exaggeration to say the cutter had a weak and dysfunctional chiefs' mess. Most of the chiefs were avoiding the mess and the command chief had basically conceded control to "Chief Personality." And he was running the mess like his own personal apartment! And the enlisted crew got a front row seat to a dysfunctional chiefs' mess and a chief and ensign friendlier then they should be.

How did it get so far out of control? There's plenty of blame to go around, but I'm just going to address the chiefs. The command chief showing a lack of leadership in the mess. The ensign's division chief not mentoring her. Chief Personality not respecting his fellow chiefs. All the other chiefs for not stepping forward and taking charge of the situation.

So what could have been done to resolve this? The command chief obviously should have taken control of his mess and had Chief Personality stand down. The ensign's division chief should have been talking some sense into her. If nothing else, one of the other chiefs should have stepped up and taken control of the mess. How do they do that if the command chief isn't willing?

They go to the commanding officer and ask for him to be replaced. Because the command chief is in that position at the pleasure of the commanding officer. If that doesn't work you go to the next command chief in the chain of command; usually at district or area level.

What makes this even worse was that after the chief and ensign transferred ashore he divorced his wife and started having a relationship with her. So you have to ask yourself, when did the inappropriate relationship start?

And this is how the respect, built up over many years, for the position of the chief gets whittled away. Because everyone who saw this play out lost some respect for chiefs and the chiefs' mess.

This is a good time to get into who I recommend you should and shouldn't be hanging out with. When I was stationed on a cutter I lived aboard. But just because I was onboard more than normal doesn't mean that I was hanging around with everyone after the work day. I relaxed in the chiefs' mess, worked in the engine room, or relaxed in my stateroom. In general I kept to myself. Sure I saw a lot of the crew in the chow line or walking around the passageways but the encounters were brief and professional.

The people I hung around with were my fellow chiefs. I didn't hang around officers and I didn't hang around junior enlisted. If I ran into officers or junior enlisted out on liberty, like in a port call, I made the encounter as brief as possible and was on my way.

Now there were the times when the wardroom and the chiefs' mess went on joint outings, coordinated by the commanding officer, and of course I went on those. And when we had an initiation we invited some of the warrant and commissioned officers who were prior chiefs. These were rare but traditional occasions.

And I know there are a lot of small units in the Coast Guard, where there is just you and a junior officer or senior petty officer. You have to make the call on how you want to interact. I would recommend trying it my way first. You can always loosen up a little if you think it appropriate. But it's really tough to go the other way; from buddies to superior/subordinate.

I just feel that nothing good is going to come out of hanging around them. I found that if you want to be a credible and impartial chief then you need to hang out with your fellow chiefs or stay to yourself. Things just have a way of working out better when the chiefs' mess, wardroom and, junior enlisted revolve within their own orbits.

One of the best bits of advice I received when I was preparing to make chief was *not to get on a first name basis with my subordinates*. And in the Coast Guard, where you have very small units, this can be challenging. But this proved to be very sound advice.

At just about every unit I went to as a chief and above I was encouraged by my subordinates to get on a first name basis. And I'm the first to admit that it's not easy, and sometimes uncomfortable, but you have to stand your ground. And I'm talking on and off the job. And there are several reasons for this.

First, you don't want to show favoritism. On an afloat unit I had 10 or more E-2 through E-6's working for me. Out of my two or three E-6's I would pick one as my leading E-6. Not necessarily the one with the most seniority, but the one that was the most competent. They would be left in charge of the engine room on numerous occasions while I attended to collateral duties. We spent a lot of time working together and the leading E-6 would eventually try to get on a first name basis. Especially if they were older than me. But I didn't let it happen and it saved me a lot of headaches. Just put yourself in their shoes. If I was one of the other E-6's, I would feel like I should also be on a first name basis with the chief. Where do you draw the line?

Second, you're more than likely going to have to pull rank or discipline the member at some point in your tour. No matter what you may think, you will always have to get firm with every one of your subordinates. I don't care if you're at a shore command working 9-5 or getting underway and spending every hour with them for 3 months. You're going to butt heads. And when you do, you want the power and the authority of your rank.

Mentor your people

In order to mentor your subordinates you need to know them. Their strengths, weaknesses, and a bit about their personal lives. The last thing you want to happen is have your E-4 come and tell you his wife was in a car accident and you didn't even know he was married. Or call your E-3's parents and find out they were a problem child. Learned that one the hard way!

You have to be a little more diligent about this at shore units because you're not interacting with your people as much as you would afloat. You and your subordinates go home in the evening and they're off your radar. When you're at work you have a tendency to levitate away from them as your leading petty officer interacts with them more. So you really have to get up and out on shore units and make an effort to "work the crowd" as they say. Because you have to know the people you're responsible for.

And this is where I really feel that mentoring pays off, especially for people who don't absorb information well from a traditional leadership course. Your subordinates can get tailored leadership training from you every day of the year. And I know what you're thinking; I don't have time to mentor my people every day. Sure you do! Just by the way you act and interact with your people you mentor them. And what do I mean by that? When I was a junior enlisted I had a chief who would come down and work in the engine room with us. While we never had personal conversations, I felt he cared about me and the engine room because he was there, working like me. And I also had the experience of working for a chief who I never once saw working in the engine room. And I never felt he cared about me, or the engine room for that matter.

I always liked this quote from Charles Schwab, "*I consider my ability to arouse enthusiasm among my people the greatest asset I possess, and the way to develop the best that is in a person is by appreciation and encouragement.*"[7]

So here's something to try. When you start your day off take a few minutes to walk around and check in with your people. You don't have to say anything, your presence will be noticed. But just asking "Hey what are you working on" in an interested tone of voice shows appreciation and encouragement. I always tried to pick when my guys were working on the most mundane task, like wiping up the bilge, to walk around and check on them. Just when they thought they were doing the most meaningless job and no one cared. I wanted to make sure they were appreciated.

Knowing your people will also help to determine how you can guide them in their career. You need to know if non-rated E-3 Smith wants to go to an aviation rating and if non-rated E-2 Jones wants to go into operations. That way you can tailor their work load, when you can, to let them experience the rating they're interested in.

Some people are hard chargers and will be submitting application for all kinds of things. But there are also the meek or shy folks who are going to be below your radar. Don't let them fall by the wayside because they're not the "squeaky wheel." I'm just going to mention a few things here so the following isn't all inclusive. But it will give you an idea.

There are so many programs out there that you should be on the lookout for. Advanced education programs, direct commission officer (DCO), and officer candidate school (OCS), just to name a few. If you have qualified candidates working for you it should be a priority for you to at least make them aware of these programs. Junior petty officers don't spend a lot of time reading Coast Guard or Navy message boards. Why? Because they're usually too busy working, too concerned with liberty, and most of all don't know about these programs.

I was always keeping an eye on my calendar for when the usual programs would roll around, as well as the message board for anything unusual. On one particular occasion I saw an out of the ordinary solicitation for afloat units. They were looking for junior enlisted to attend a USO celebration in Germany. I had an exceptional E-4 and put him in for it. He was selected and got to represent the Coast Guard with an all-expense paid trip to Germany!

Push your people to get their degree. A good time to address this is when you're going over their marks with them. It's the first term enlistees who typically don't consider going to school. And they fall into one of three categories. First you have the folks that plan on being one term enlistees and using the G.I. Bill when they get out. Second, are the folks who plan to make the Coast Guard a career, but don't believe they can fit schooling into their already busy work lives. And last, there are those that are just not interested in higher level education. For the first group, I always told them that life comes at you fast. You just may decide to stay in. So why not work on your degree while you're active duty? Then, if they did get out, they would be ahead of the game. To the last two groups, I liked to ask them questions to try and find out why they felt that they didn't have time or why they had no interest. From there I adjusted my mentoring. Some people really are just plain not interest. I know, I was one of them. I didn't start to pursue my degree until I was a senior chief.

But as our society has become better educated and more people have their degree, it has become the standard.

So, for the time being, at least in the military services, if you want your people to be competitive for any type of special program they are going to need advanced education to qualify. So be an advocate for education!

If nothing else, your people should be working on their technical certifications. When you're active duty you really don't follow what's going on in the civilian sector of your specialty. Therefore, when you get out and try to get a job you're surprised to learn that your years of military service don't equate to civilian certification; such as apprentice, journeyman or master, because the military doesn't certify that way. The military does have a program that allows you to keep track of your technical progress so that when you get out it's transferable to civilian certification. But this has to be compiled as you progress through your military career and just can't be thrown together a few months before you're discharged. So make sure you understand the program and educate your folks on it.

As I said before, some people have no business advancing. Either they are not technically qualified or they are not yet mature enough. This is where the wardroom is counting on your expertise in knowing there capabilities and potential. Because not everyone should be promoted to a position where they may play a role in other people's careers when they're not capable of leading.

This is one of those times when being a chief will be tough for some people. Because you have to be able to look the junior petty officer in the eye and say "You're not ready for advancement." Nobody wants to be the bad guy, and how do you know if you don't give them a chance? That's the big question. And that's why you're a chief with all the special privileges, making the big bucks. You should know your people well enough to know whether they have a shot at it or not. If you've been mentoring them, and working alongside them, and communicating with them, your decision will not be a surprise. And their performance reports/marks should be reflective of your positive or negative recommendation.

And I'll close this chapter with my thoughts on the evaluation, or marks, system.

This is a great opportunity for you to engage your subordinates and help them maintain realistic goals based on their *actual* performance.

But it has a huge potential flaw - the supervisor. Supervisors have a tendency to either not care enough to take the time nor make the effort required to document high or low performance. Or they exaggerate performance, trying to give their subordinate an advantage on a promotion or board.

Don't perpetuate this. All you're doing is delegitimizing the evaluation system and confusing your subordinate with an inaccurate evaluation. Marks are important and deserve your time and effort. Mark your people per the evaluation guidelines and the individual's *actual* performance. They will do just fine with the fair and accurate marks you give them.

Life after Chief

Even though making chief petty officer is a major milestone in your career, it shouldn't end there. Along with "words of wisdom" requests I'm still asked for advice on career paths - stay enlisted or become an officer. And I'm asked this because I served as a master chief and commissioned officer, so I know both sides.

Becoming a commissioned officer was a great career decision so obviously I'm going to be a little biased and recommend becoming an officer. And when making the decision to become an officer it should come with having a long range goal of staying in the military. But let me lay out my case, and you decide.

First my background and thought process on becoming an officer, because some of my concerns and hurtles may also be yours. I had been a master chief before taking my commission as a lieutenant junior grade. And I had an opportunity to become a warrant officer when I was a senior chief, but declined the offer because I enjoyed working in my rating.

I thought if I stayed enlisted I could still enjoy hands on involvement with my specialty. The sad fact is that when you get into the "super chief" ranks – senior and master chief – you're expected to move out of your rating specialty and into general leadership and management roles. That's a big reason why the service wide exams for these ranks are heavier on general military knowledge then technical specialty.

And I found this out the hard way. As a senior chief I should have seen the writing on the wall when I filled a non-technical white collar billet, which I hated. After taking a subsequent assignment afloat I found happiness again in my rating. Then when I made master chief I was lucky enough to slide into another technical job. But I could see that they were limited. There were only 33 master chiefs in my rating - one of the biggest ratings in the Coast Guard - and only a few of these jobs were in my technical specialty. Most were leadership and management positions. So I had to make a decision; stay in the world of these 33 jobs or expand my options to thousands of jobs. And to top it off, due to high year of tenure, I only had about 7 more years before I would be forced to retire.

Whether I stayed on the enlisted career path or became an officer, I was going to be filling mostly white collar positions. And if I was basically doing the same thing either way, why not choose the path that paid more, offered more job flexibility, and extended my high year of tenure. In this case, my high year of tenure clock was reset, letting me basically stay in until age 62. And in hindsight, it was the best decision for me at the time because my retirement check as a lieutenant is much more than I would have received as a master chief.

Your promotion potential as an officer is greater. In the enlisted career path you have to compete for an ever shrinking pool of jobs. Senior chief is only 3% and master chief a mere 1% of the enlisted workforce. And these limits are set by law. There are no warrant officer manning limits. And as long as your performance evaluations are satisfactory you're promoted to the next higher paygrade automatically when you reach your time in grade milestones. And the warrant officer option still keeps you for the most part in your technical field; although a white collar supervisor.

As a commissioned officer you face a controlled number of billets but there are thousands of them compared to the "super chiefs" and warrant officers. Promotions are not guaranteed but you have a good chance of obtaining lieutenant commander or higher before you're forced to retire. And your high year tenure clock is reset. So you can pretty much stay until the mandatory retirement age of 62. That was a big issue I ran into with my fellow master chiefs. At my last assignment, as a lieutenant, I supervised eight master chiefs who were the Coast Guard's Rating Force Master

Chiefs. And most of them were *forced* out at their 30 year mark due to high year of tenure.

If I had stayed a master chief I could have been one of those forced out. One of the jobs in my pool of 33 billets was one of these rating force master chiefs. And I could have been working for a lieutenant instead of being the lieutenant.

Now with all that said I recommend applying for the warrant officer program if you have under 20 years of service, because this will allow you to retire as a W-4. You're basically guaranteed promotion through the warrant officer ranks. You can still pretty much stay in your technical specialty. And W-4 pay is comparable to lieutenant commander, much more than a master chief. And as a warrant officer you have a direct path applying for a commission.

If you have over 20 years of service the commissioned officer path is the better option, because your high year of tenure clock resets. This gives you the opportunity to stay in long enough to possibly make lieutenant commander or higher. Plus there is the added benefit of having the most job options to choose from.

Whichever option you choose – staying enlisted or becoming an officer - always plan for your future. Most people assume that they'll retire when they're eligible, around their 20 year mark. Yet many of these same people go a lot longer, into 25 to 30 years. Why?

Because they find job satisfaction in white collar leadership and management positions, like the money, and find they enjoy being in the military. So don't sell yourself short. What you want today as a new chief with 15 years of service may not be what you want as a senior chief with 22 years. So stay *flexible* and *promotable*!

Chiefs Creed

Usually after the chief's initiation your copy of the Chiefs Creed gets folded away with your Chief's Charge Book; both never to be seen again. But remember how you worked so diligently going around to all those chiefs seeking "words of wisdom?" Did you even read what was provided to you? If not, you're doing yourself a great injustice. You were provided with some valuable advice and guidance. Go through them and pick out the ones that you feel are valuable, write them down, and keep them handy. When you have challenging days pull them out and read them, they may just have the answer you're looking for. If nothing else you'll feel better about your situation when you read that you're not the first chief to have problems.

I also encourage you to read your Chief's Creed every now and then. A good time is just before you report to your next unit. It will help you refocus your thoughts on your position and rank. And when you read it, think about what the Creed is conveying to you.

I like to focus on this phrase;

"The exalted position you have now received, and I used the word "exalted" advisedly, exists because of the attitude, the performance of the Chiefs before you. It shall exist only so long as you and your compatriots maintain these standards."[8]

It's a very powerful statement. And it served as great encouragement to me as I prepared to check into each new unit. In my own words, I would think of it this way:

"The power of the anchors I'm wearing when I cross the quarterdeck is derived from the chiefs before me. And that power lasts only as long as I maintain their high standards. I'm given a huge advantage when I walk across that quarterdeck; the legacy of the chief. So don't squander that advantage away."

I've laid out the Creed here for you on the following pages as an easy reference. It's the one from when I was initiated in 1994.

Chief Petty Officer Charge[8]

During the course of this day you have been caused to suffer indignities, to experience humiliation. This you have accomplished with rare good grace, and therefore we now believe it fitting to explain to you why this was done. There was no intent, no desire to insult you, to demean you. Pointless as it may seemed to you, there was a valid, time honored reason behind every single deed, behind each pointed barb.

By experience, by performance and by testing, you have been advanced to Chief Petty Officer. You have one more hurdle to overcome. In the United States Coast Guard, E-7 carries unique responsibilities. No other armed forces throughout the world carries the responsibilities nor grants the privileges to its enlisted personnel comparable to the privileges and responsibilities you are now bound to receive and are expected to fulfill.

Your entire way of life has now been changed. More will be expected of you, more will be demanded of you. Not because you are an E-7, but because you are now a Chief Petty Officer. You have not merely been promoted one pay grade, you have joined an exclusive fraternity, and as in all fraternities, you have a special responsibility to your fellow Chiefs, even as they have a special responsibility to you.

These privileges, these responsibilities, do not appear in print. They have no official standing, they cannot be referred to by name, number or file. They exist because for over 200 years, Chiefs before you have freely accepted responsibility beyond the call of printed assignment. Their actions and their performance demanded the respect of their seniors as well as their juniors. It is now required that you be the fount of wisdom, the ambassador of good will, the authority in personnel relations, as well as technical applications. "Ask the Chief" is a household word in and out of the Coast Guard. You are now the "Chief."

The exalted position you have now received, and I used the word "exalted" advisedly, exists because of the attitude, the performance of the Chiefs before you. It shall exist only so long as you and your compatriots maintain these standards.

So this then is why you were caused to experience these things. You were subjected to humiliation to prove to you that humility is a good, a great, and necessary change, which cannot harm you. This, in fact strengthens you, and in your future as a Chief Petty Officer you will be caused to suffer indignities, and to experience humiliation far beyond those imposed upon you today. Bear them with the dignity, with the same good grace with which you bore these today.

It is our intention that you will never forget this day. It was our intention to test you, to try you, to accept you.

Your performance has assured us that you will wear your hat with pride of those before you.

We take a deep sincere pleasure in clasping your hand and accepting you into our midst.

References

1. "Twelve O'clock High" screenplay by Henry King, Sy Bartlett, and Beirne Lay Jr.

2. "The Caine Mutiny" screenplay by Stanley Roberts

3. U.S. Navy Judge Advocate General Corps Results of Trial reports from January through December 2016

4. Navy Times article dated September 2015

5. U.S. Coast Guard Good Order and Discipline reports for the 1st through 4th quarters of FY 2016

6. U.S. Coast Guard Good Order and Discipline report for the 3rd and 4th quarters of FY 2016

7. Dale Carnegie "How to Win Friends and Influence People" copyright 1936, 1964, and 1981

8. U.S. Coast Guard Chief Petty Officers Association

Notes

About the author

Ed Semler retired from the United States Coast Guard in December of 2007 with over 25 years of military service in both the United States Army and United States Coast Guard. In the Army he was an enlisted combat bridge crewman (12C) and was honorably discharged as a specialist four (E-4). While in the Coast Guard he was an enlisted machinery technician (MK) obtaining the rank of master chief petty officer (E-9), was commissioned as an officer, and retired as a lieutenant (O-3E).

After his military career Ed dabbled in teaching at a Vocational Technical School and was a self-employed plumber for several years. As a pass time he enjoys reading and writing.

His other publications are;

"Around The World," which details his 25 years of service as an officer and enlisted man in the U.S. Army and U.S. Coast Guard.

"U.S. Coast Guard Cutter Sherman (WHEC-720) Circumnavigation Deployment 2001," which details the

Sherman's historic circumnavigation of the globe and deployment to the Persian Gulf in 2001.

"*The Three Gunsallus Brothers*" which details three brothers fighting for the state of Pennsylvania during the Civil War.

Fully retired he resides in Schulenburg, Texas with his wife Jana, a retired Air Force Senior Master Sergeant. Please feel free to contact him at mkcm378@gmail.com check out his website www.edsemler.com or his YouTube channel www.youtube.com/@MKCMLT

www.ingramcontent.com/pod-product-compliance
Lightning Source LLC
Chambersburg PA
CBHW061336040426
42444CB00011B/2953